The First Travel Guide to the Bottom of the Sea

The First Travel Guide to the Bottom of the Sea

by RHODA BLUMBERG

illustrations by GEN SHIMADA

LOTHROP, LEE & SHEPARD BOOKS   NEW YORK

For Jerry, who shares my trips.

Text copyright © 1983 by Rhoda Blumberg
Illustrations copyright © 1983 by Gen Shimada
All rights reserved. No part of this book may be reproduced or utilized in any form or by any means, electronic or mechanical, including photocopying, recording or by any information storage and retrieval system, without permission in writing from the Publisher. Inquiries should be addressed to Lothrop, Lee & Shepard Books, a division of William Morrow & Company, Inc., 105 Madison Avenue, New York, New York 10016. Printed in the United States of America. First Edition.
1 2 3 4 5 6 7 8 9 10

Library of Congress Cataloging in Publication Data

Blumberg, Rhoda.
  The first travel guide to the bottom of the sea.

  Bibliography: p.
  Summary: Describes an imaginary voyage of the submarine Sea Dragon to the ocean bottom, where many strange sea creatures are observed in their natural environment.
  1. Marine biology—Atlantic Ocean—Juvenile literature. 2. Atlantic Ocean—Juvenile literature. [1. Marine biology. 2. Atlantic Ocean. 3. Ocean bottom] I. Shimada, Gen, ill. II. Title.
QH92.B58   1983   574.92'1   82-17938
ISBN 0-688-01692-8

CONTENTS

NOTE FROM THE AUTHOR

I used to believe that the bottom of the ocean was as flat as a pancake and as devoid of life as the Moon. When I began reading about the wildlife that live miles beneath the waves, and when I learned that huge mountain ranges can be found underwater, I knew that I had to see the sea with my own eyes. Therefore, I dreamed up the Super Submarine *Sea Dragon*. It is a fantasy ship that allows us to explore the Inner Space of the ocean.

Although the submarine is make-believe, the sights that one sees outside its windows are real. Fish doctors' offices, lilliputian colonies, and trenches and cliffs may sound fantastic, but these truly exist.

I would like to thank Dr. C. Lavett Smith, Curator of Ichthyology at the American Museum of Natural History. He guided me through untroubled waters and made ocean adventure a joy. He made suggestions, read my manuscript, and verified facts. Planning a voyage on the *Sea Dragon* was a delight for me. I am in the ship's hold, reluctant to disembark.

I hope that the day will come when large nuclear submarines will not be loaded with missiles for war. Instead, I dream of staterooms for all of us, and wide windows that will reveal the marvels of seascapes and of ocean creatures.

Rhoda Blumberg
Faraway Farm

I. All Aboard

Welcome aboard the Super Submarine *Sea Dragon*. You are embarking on an in-depth adventure far more fascinating than any trip to the Moon. While space probes are traveling billions of miles to discover if there is one-celled blob of life in Outer Space, you will be exploring Inner Space, where you will see many weird and wonderful beings.

You will be invading a Liquid Land of Odd. Some creatures have fifty arms; others a hundred eyes. You may see a whale whose tongue is as big as a baby elephant and whose heart is the size of a motorcycle. And you'll undoubtedly watch mini-monsters that flash lights built into their bodies.

You will experience close encounters of a thousand kinds. Queer fish and monstrous-looking animals will swim and drift past windows. While some inhabitants of the watery world are interested in examining you, others will be frightened and attack. Don't be surprised if a swordfish tries to stab your ship, a whale shoves your sub, or a squid wraps itself around one of the *Sea Dragon*'s outside lights. Keep in mind that your submarine is a UFO invading their world. You are the alien from a strange, way-out land.

Sight-See Inside the Sea

We live on the only liquid planet in the entire solar system. More than 70 percent of our Earth is under water. *The highest mountains, the largest mountain ranges, the sharpest cliffs, and the flattest plains are fathoms below. There are huge valleys, and canyons deeper and steeper than any on dry land.*

A group of fishing, mining, and oil corporations built the S.S. *Sea Dragon*. (S.S. in this case stands for Super Submarine.) It was originally intended exclusively for scientists and technicians. To add income and to increase public interest in their investments, the corporations converted their Super Sub into a luxury liner that accommodates one hundred tourists. They still use the *Sea Dragon* for research.

Experts in ocean life and underwater geography are members of the crew. They will guide you to the bottom of the sea.

Officers and Members of the Crew

Captain Hugo Down is in charge. This seasoned old salt was once a junior officer on the *Nautilus*, the world's first nuclear-powered submarine. Built by the United States Navy in 1954, the *Nautilus* sailed under the North Pole in 1958, traveling more than four-hundred feet below the ice. Before reaching the Pole the ship sailed more than half a mile above a ten-thousand-foot-high mountain range called the Lomonosov Ridge. The mountains run across the bottom of the Arctic Ocean in almost a straight line for more than a thousand miles. Over mountains and under ice! What an inspirational, uplifting experience for young officer Down!

Staff Captain Luke Spywater is second in command. Spywater assigns duties to the crew. He keeps examining every part of the sub to make sure it's shipshape. No risk of a leak with Spywater aboard!

Like Captain Down, Spywater is an old-timer. In 1960 he sailed aboard the U.S. Navy's nuclear submarine *Triton* when it followed the route taken by Magellan's ships in 1522. Around the world in eighty-four days: a journey of 36,014 miles in Liquid Space before the *Triton* surfaced. What a breathtaking experience!

Chief Engineer Sally Forth is in charge of maintenance and repairs. Last year she was executive engineer for the California Underwater Tractor Company. Farmers who lease plots of undersea land from the government buy the tractors. They harvest giant kelp, a type of seaweed found in west coast waters. It's harvested all year round. Because the submarine tractor doesn't pull out the roots, farmers don't have to seed the sea.

Kelp farming is big business. It's marketed as food

and as fertilizer. You have eaten a product called *algin* made from kelp. It's in commercial ice creams, puddings, salad dressings, jellies, jams, and medicines. Some kelp are as tall as an eighteen-story building, and grow at the rate of two feet a day.

Purser Buck Moneysworth is in charge of Davy Jones' Locker, space used to store money and valuables. You have no use for cash while you're cruising under the ocean. Buck will lock it up and shell it out when you reach port.

Purser Moneysworth used to pan for gold in the streams of California. Buck still has a glint of gold in his eyes. He knows that ancient streambeds in the ocean contain gold and that other precious metals can be found there, too. He hopes that as a result of this job he'll impress some bigshot mining executive and be hired as an undersea miner.

Chef Saki Yaki is chief officer of the kitchen. How fortunate that the *Sea Dragon* was able to recruit this world-famous chef! He attended a school in Tokyo that teaches cooks the art of making fugu stew, a delicacy made from poisonous puffer fish. Graduates receive licenses from the Japanese government. Only a master chef is qualified to remove the deadly venom from a potted puffer. A few honorable graduate fugu stew cooks have been careless, and their customers found that death was but a swallow away. Chef Saki Yaki has had a blameless career, so far.

Note: No fugu stew is served on board.

The main food supply grows, hops, swims, drifts, and crawls outside the ship. *Yul Nibble*, Yaki's assistant, uses a metal claw attached to a vacuum hose called a *slurper* to bring sea plants and ocean creatures into the galley. Both the claw and the slurper are controlled by knobs inside the kitchen.

Saki Yaki examines each specimen. Although there's no worry about fish being fresh, he must be sure that a poisonous animal or plant doesn't end up in the pot.

Chef Yaki constantly experiments with new dishes. Rattails, for example, among the most plentiful fishes in the ocean, have never been caught commercially because people have been turned off by the name. Saki Yaki is smart enough to call rattail Emperor's Delight. It has become one of his most popular dishes.

Yaki is performing a useful service. He is aware that out of over 25,000 species of fish only a handful are caught for food. People of the world go hungry, even though the sea teems with delicious, protein-rich plants and animals.

Some of Saki Yaki's other specialties: urchin soup, fish eggs fondue, French fried small fry, plankton pancakes, seaweed salad, stuffed beak of squid, flounder fish sticks, and octopus (cooked for eight).

A variety of frozen foods is available for those who don't want seafood.

II. Tour of the Ship

It's down the hatch and into the belly of the *Sea Dragon*. A spiral staircase goes below to Levels A and B. All tourists' bunks and baths are located one flight down, on Level A. Ship stewards help you find your room. Sorry, no portholes in sleeping quarters. The adhesive used for inserting each pane of glass is astronomically expensive. You will watch wildlife through huge glass windows in other parts of the sub.

After you have placed your belongings in the locker under your bunk bed, go topside. It's exciting to be on deck as the ship takes off from Norfolk, Virginia, for a three-week holiday.

Before the *Sea Dragon* sinks, Captain Spywater always takes passengers on tours of the ocean liner. He conducts four of them so that no more than twenty-five people are in each group.

Topside

You inspect the conning tower on Top Deck. The conning tower looks like a giant glass bubble. When the ship is on the surface, it is used as a lookout by two crew people on watch.

Their eyes are glued to binoculars as they try to sight unusual sea creatures that might be leaping out of the water or riding the waves.

Lean over the railing and look at the sides of the ship. Captain Spywater points out some equipment attached to the hull. You see still and motion picture cameras that take underwater photos. Many of them film scenes that will appear on television screens inside the sub. You also see the *Sea Dragon*'s four claws: one for the kitchen chef; the others for scientists who gather rocks, mud, fossils, plants, and wildlife specimens. Slurpers are attached to each claw. Superpowered lights rim the hull. They're often switched on when the ship goes under so that you can watch ocean life.

The hull is made of the world's strongest material—a secret mixture of various metals developed especially for the *Sea Dragon*. It must be able to undergo extreme pressure. The weight of water miles down pushing against ordinary steel would crush a ship the way a steamroller can flatten a paper cup. Water weighs almost eight hundred times as much as air and increases in pressure as you descend.

The *Sea Dragon* can keep in shape five miles beneath the sea, where the pressure is 800 times more than land atmosphere.

UNDERWATER PRESSURES

On dry land	= 14.7 pounds per square inch
1 mile down	= 2,352 pounds per square inch
2 miles down	= 4,704 pounds per square inch
3 miles down	= 7,056 pounds per square inch
4 miles down	= 9,408 pounds per square inch
5 miles down	= 11,760 pounds per square inch

The air pressure is perfect. Just as the cabin of a jet plane maintains normal pressure when it's miles up, so the air inside the *Sea Dragon* is adjusted. If this weren't so, your lungs would collapse like pricked balloons. Oxygen is extracted from the surrounding seawater. The carbon dioxide you breathe out is absorbed by chemicals in the walls.

The Nose

Follow Spywater to a room up front called The Dragon's Nose. The Nose, a semicircle of glass, is the main observation lounge. *Because glass becomes stronger under pressure, there's no chance of it cracking up.*

You'll undoubtedly see hammerhead sharks and barracudas eyeing you from the other side of the glass pane. They habitually swim alongside the *Sea Dragon* as soon as it enters southern waters.

S.S. SEA DRAGON

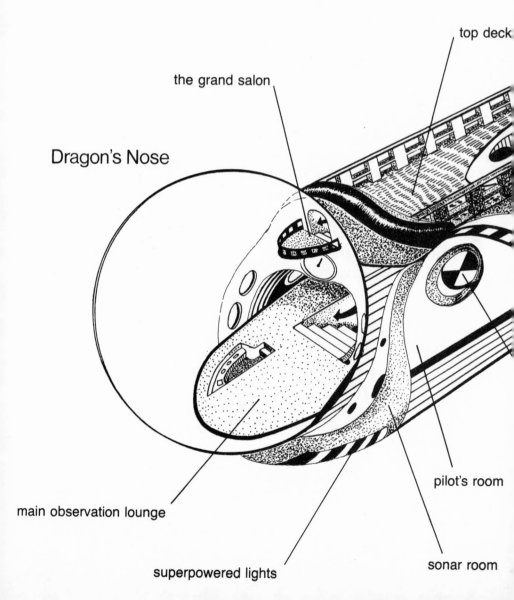

top deck

the grand salon

Dragon's Nose

main observation lounge

superpowered lights

pilot's room

sonar room

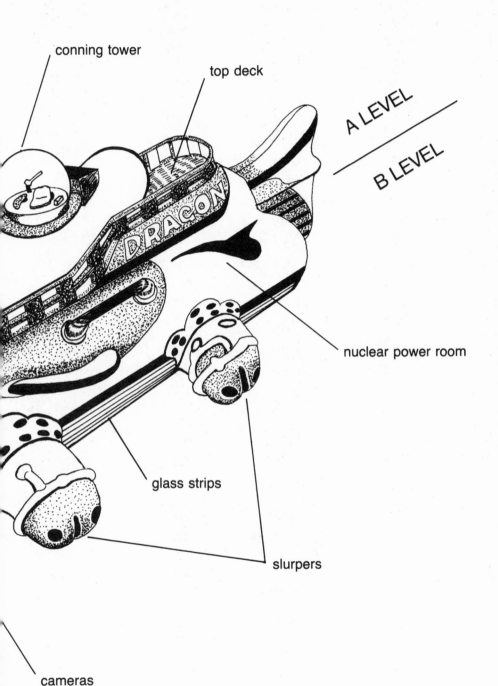

conning tower

top deck

A LEVEL

B LEVEL

DRAGON

nuclear power room

glass strips

slurpers

cameras

You'll be called to the Nose whenever something interesting takes place outside. There's a speaker in every room. Even when it's in the middle of the night, a mate on watch announces a special happening that is taking place in the ocean. Put on a robe and hurry! (Turn off your speaker if you don't want to be disturbed. But you'll be sorry. You won't have any fish stories to tell at breakfast.)

The Floor of the Sub

When there's activity *underneath* the ship, you'll be told to go down to Level B. Strips of glass on the floor of the sub enable you to see what's going on below.

Sucker fish often cling to the glass. The suction discs on top of their heads look like the bottoms of sneakers. Known as hitchhikers of the sea, these fish usually attach themselves to sharks, whales, and other big ocean dwellers. Suckers don't stay long. They must swim off to hunt for food.

The Engine Room

The ship's engine is also on Bottom Level. It is powered by a *fusion reactor* that takes hydrogen from sea water and heats it until the hydrogen atoms are fused (forced together). This produces atomic power, which runs the ship.

No need for thousands of gallons of fuel. Nor are radioactive fuels like uranium necessary. (Uranium is used on other nuclear submarines.) The *Sea Dragon* is the only ship in the world with a fusion reac-

tor. It is fast and quiet and the power could keep it going forever.

You see a brilliant light that glows from the super-hot reactor engine.

The Sonar Room

Sonar means echo locating. High-pitched sounds sent into the water come back as echoes that bounce off cliffs, canyons, other ships, and the sea floor. The volume and length of time it takes sounds to return indicate size, shape, and distance of surrounding objects. These are recorded on screens. Images on two screens give pilots an exact picture of the seascape. In addition to screens that light up, another screen has a pen that traces a pattern over moving graph paper. The tracing is an exact map of the ocean floor.

Scientists are now studying members of the whale family in order to improve sonar equipment. Experiments conducted in aquariums have demonstrated that blindfolded dolphins are able to locate a fish that is only three-quarters of an inch long. In complete darkness, dolphins avoid wires and nets. It takes about one ten-thousandth of a second for echoes from an object a foot away to come back and be heard by these creatures.

The *Dragon*'s sonar causes echoes to bounce off shoals of fish, sometimes indicating the number of animals in them. Listen to tapes that record conversations between Roger Beep, head of the Sonar Room, and Captain Down. Here's a sample:

"Captain . . . this is Beep! Approximately five hundred fish—they may be bluefin tunas migrating north—all eight feet long—coming around the mountain."

"Roger. Go over the mountain. Get as close as you can, so that passengers can see, but don't disturb school."

The Pilot's Room

The Pilot's Room, adjoining the Sonar Room, is a jumble of dials and screens. A small computer guides the pilot.

Television cameras on the hull act like eyes as they fill screens with images of the immediate surroundings. The room has a big compass on a stand, and a chart table with ocean maps and the ship's log. No picturesque skipper at the helm. Instead, pilot and copilot sit in comfortable lounge chairs turning knobs and pushing buttons as they ride the *Dragon*.

The Gift Shop

The tour ends on Level B. A small gift shop sells rare shells, fossil bones, and unusual fish specimens that have been collected by the ship's mechanical claws. Sharks' teeth and whales' earbones that literally litter the ocean floor are inexpensive collectors' items. They make wonderful conversation pieces for home whatnot cabinets.

Many specimens of sea creatures aren't for sale. They are sent to scientists. Valuable medicines can be taken from the bodies of sea creatures. For example, certain sea snails provide a drug that relaxes muscles, and certain octopuses have a chemical that reduces blood pressure. New species that are slurped in every year are treasured by marine biologists, who know that many important secrets lie locked inside ocean dwellers.

The Grand Salon

The Grand Salon takes up most of Level B. The Salon is a combination dining-recreation room. A balcony into the Dragon's Nose reminds you that you're inside Earth's aquarium.

Video games, bingo, backgammon, and cards are played here. You are allowed to gamble if you use sea shells, sand dollars, or sharks' teeth instead of money.

Exercise equipment is at one end of the Salon. Treadmills, rowing machines, push-and-pull devices, and stationary bicycles enable you to keep in shape. Ultraviolet lamps should be used to make up for the lack of sunshine.

Tony Chestnut, physical education counselor, supervises the equipment and conducts exercise classes.

In addition to a library of books and taped concerts, the Salon features its own musical instruments. An old-fashioned upright piano that's over one hundred years old has keys made of baleen. Baleen is a horny bristle that grows inside the mouths of baleen whales. (It was used to stiffen your great-grandmother's corsets.) The piano stool, with its seat of woven baleen and its pedestal made of a whale's backbone, is also antique. Fortunately, oddities of this sort are no longer made. Because of the work of conservation groups, whaling is no longer a big industry.

Conch shells, which once housed mollusks, have holes carved into their sides. These make fine trumpets. A Florida horse conch, two feet long, is the loudest instrument in the band.

Passengers are welcome to play during "Noisy Hour," from five to seven P.M. Occasionally, members of the crew who call themselves the "Marine Band of Blowhards" give concerts. Should you want to tune them out, clamp on earphones and listen to lectures or concerts on cassettes.

Evening events scheduled for the Grand Salon are listed in *The Spitfire*, the *Sea Dragon's* weekly newspaper. There are movies, talent shows, and a weekly quiz show called "Name that Fish." Marvelous lectures take place. Cliff Rockbottom, the famous geologist, is aboard. He's ever ready to tell an audience about the latest discoveries in the hills, mountains, and valleys of the ocean. Noted ichthyologist (fish expert) Gil Haddock is also a scheduled speaker on every cruise. Gil not only knows fish, but mollusks and echinoderms, too. (Mollusks are soft-bodied animals. They include snails, clams, octopuses, and squid. Echinoderms are marine animals like starfish and sea urchins.)

At the end of your tour you are invited to drink delicious Dragon Punch, the traditional welcome-aboard refreshment. The punch is served from a giant clam shell that once housed a three-hundred-pound blue clam that had looked at the world through hundreds of eyes.

III. The Continental Shelf

Three hours after leaving Norfolk, the *Sea Dragon* sinks under the waves and cruises above the *continental shelf*. The shelf is the shallowest part of the ocean floor. All continents have shelves rimming them. They are rarely deeper than six-hundred feet, and average between two-hundred and three-hundred feet. Millions of years ago these shelves were dry land. Need proof? Look at the gift shop's display case. It features elephants' teeth and the bones of the following animals: a giant sloth, a giant moose, a musk ox, and a horse. These fossils were dredged from the sea floor below you. Rare specimens: not for sale.

Go to the Nose to fish-watch. Most of the fish we eat are caught above continental shelves. At times so many swim by it's hard to see the oil rigs and wrecked ships that are scattered along the coast's sea floor. You pass ships that were beaten in battles, or battered by storms. As the *Dragon* heads for the waters of Cape Hatteras you see fewer rigs and more wrecks. There are old wooden ships that have been broken to bits, and iron hulks that are covered with rust. Most wrecks, half buried in the ocean floor, look more like trash than the proud ships that once sailed the seas.

Take heart! Look down! You are about to see a sunken vessel that is listed in the United States National Register of Historic Places.

A Famous Wreck

The *Monitor* of Civil War fame is a national treasure. It lies thirty fathoms (180 feet) below, twenty-five miles south of Cape Hatteras. The *Monitor* is now a prime tourist attraction. How ironic that this important ironclad ship was once described as "a cheesebox on a raft" and "a tin can on a shingle." Its battle against another ironclad gunboat, the South's *Merrimac*, ended the era of wooden fighting ships.

The *Monitor*'s duel with the *Merrimac* finished in a stalemate, but less than ten months after the fight, the *Monitor* met its doom. It was defeated by the stormy waters of the Atlantic. During a gale, while being towed to port, the *Monitor*'s towline broke. Sixteen of her crew of sixty-five went down with their ship.

Lost but not forgotten! You see the *Monitor*'s shape clearly—172 feet from bow to stern.

The remains of the *Monitor* were discovered in 1973. It is so corroded and fragile that it can't be raised. Instead of being on display in a museum it is on a continental shelf, preserved in brine, nature's salt solution.

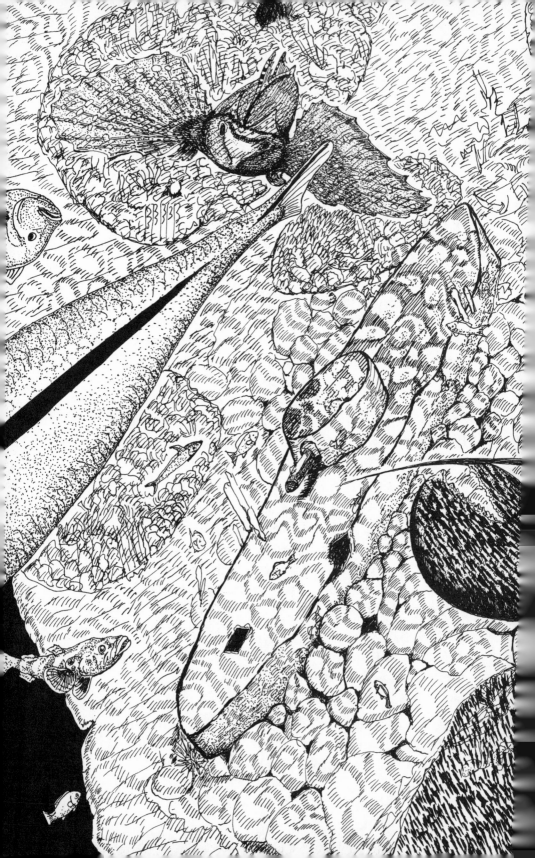

IV. Lounging and Listening While You Cruise

Beyond the continental shelf lies the deep sea, called the *abyss*. You must first go over a *continental slope* at the edge of the shelf. The slope goes down to the abyss, which in this area is more than two miles below.

It takes two days before you reach your next destination in the Caribbean. As you head south the *Dragon* travels thirty-six hundred feet (six hundred fathoms) under the waves, well below overseas naval traffic, and safely above bottom.

The Noisy Sea

You will find yellow earphones plugged into many of the submarine's lounge chairs. These are attached to *hydrophones*, microphones that have been lowered into the water. Put on one of the headsets. You will discover that the ocean can be as noisy as a tropical jungle.

You'll hear all sorts of sounds and songs. Although they have no vocal cords, fish make noises. They do this by grating teeth that grow deep in their throats; by rubbing fins against their sides; or by drumming inside muscles against their air bladders.

Groupers growl, croakers croak, haddocks hum, sea robins croon, and whales sing eerie, but enchanting, melodies. Whistles, squeaks, clicks, cackles, shrieks, and ghostly moans are picked up by hydrophones.

Sound travels five times faster in seawater than in air, and much further. Under water, sound can travel at almost a mile a second. The explosion of a one pound block of dynamite can be heard half a mile away on land. A similar explosion at mid-depth in the ocean travels thousands of miles.

Humpback whale songs are among the loudest sounds made by any animal. The creatures compose medleys of tunes that can last half an hour. Their songs change from year to year. The melodies even differ in various parts of the world. The songs of humpbacks in Bermuda waters are completely different from those of humpbacks in the balmy seas of Hawaii. Whole pods often join together in chorus. High voices, possibly of young males, join in with deep basso voices, possibly of the elders.

Perhaps they sing to attract females. It may be that they sing for their own pleasure. Scientist William Schevill taped a humpback that seemed to be amusing itself singing in a Caribbean submarine canyon. The whale went up and down the musical scale, waiting for echoes. It didn't stop until a group of whales called. Then it swam off to join them.

What is the humpback singing about? What is it saying? Will we ever understand the whale's thoughts and feelings?

Convinced that a higher intelligence could understand, scientists have placed a recording of humpback whale songs inside the Voyager Spacecraft. It is now traveling on a voyage through the galaxy that will last 1.2 billion years. Reason: The songs may be decoded by super-brainy outer space creatures. Perhaps they will understand whale talk better than English, which is also included in the Voyager payload.

Solo Sings

Mme. Mia Solo, world famous prima donna opera singer, is on board in order to test her artistry on inhabitants of the sea. This is her fifth trip on the *Sea Dragon*. The last time, she broadcast her rendition of "The Prize Song" from *Il Felice Pescatore* (The Happy Fishmonger). When her high C's vibrated inside the high seas, large schools of herring came from all directions. This may have been coincidence. Therefore, Mme. Mia Solo must repeat her experiment many times before her efforts are applauded. A special microphone has been set up for her on the starboard side of Level B.

Startling Sounds

In 1942, during World War II, the springtime mating calls of small fish alarmed Navy headquarters at Chesapeake Bay. The Navy had set up a hydrophone network in order to catch the sounds of enemy submarines. Noises like "a pneumatic drill tearing up pavement" turned out to be the love songs of 300 million croaker fish. They were drumming on their air bladders and creating a racket that lasted from sunset to midnight.

The same year this Chesapeake Bay scare took place, Navy men in the Pacific were alarmed when their hydrophones picked up crackling noises that sounded like static from a radio set. No Japanese submarine was in the area. The noises were made by a species of tiny shrimp, called snapping or pistol shrimp. Thousands of them living in holes of dead coral were clicking a knuckle-like joint on one of their claws. The clicking produced jets of water meant to scare the shrimps' enemies. Those shrimp managed to frighten one enemy: the hungry men of the U.S. Navy.

V. The Puerto Rican Trench

You travel onward and downward to a nether world where life and seascape haven't changed for millions of years.

The *Sea Dragon* sinks into the Puerto Rican Trench, seventy miles north of Puerto Rico. The trench, which is the deepest part of the Atlantic Ocean, is 450 miles long, and almost 5 miles down anywhere you measure it. Its deepest point is 30, 238 feet.

When you begin your descent you notice that the trench walls are made up of a gigantic series of naturally formed steps. After you have had a good view of these giant stairs decorated with colorful animal sponges, outside lights are switched off. The *Dragon* makes its way down guided solely by sonar.

Eerie Lights
Blinking, glowing, flashing sea creatures are outside. *Most animals living more than a thousand feet down have lamps built into their bodies.* They use

them to find their way, "talk" to each other, locate food, attract mates, lure prey, and dazzle enemies.

Hundreds, even thousands of light organs have been found on single sea creatures. The organs are located on every part of the body, even on teeth, tongues, and eyeballs. Some lights are produced by glands similar to those of fireflies. Others are made by sparkling bacteria that live in or on the bodies of ocean animals.

You may see a light brigade of tiny jellyfish that look like glowing red fireballs. Perhaps you'll sight the famous tiny Caribbean flashlight fish. Bright green lights shine from kidney-shaped pockets underneath its eyes. The pockets are filled with light-producing bacteria. Should an enemy approach, the flashlighters cover their eye pockets with lidlike folds of skin. Then presto!—they disappear in their black world.

There are tri-light fish with yellow headlights, yellow and red side lights, polka dots of yellow beams, and patches of glow-green. Hatchetfish, whose thin flat bodies have given them their name, are usually studded with blue lights. Some, however, glow ruby red.

Even more spectacular are the 170 species of lantern fish. They flash blue, green, yellow, and red. Rows of lights dotting their sides look like portholes on a toy ship. Some lantern fish have headlamps able to send beams two feet ahead. Others have wagging tongues that light up, inviting victims to enter their jaws. Viperfish also have luminous patches inside their mouths, which are never closed because their fangs are so big. Some viperfish sport whiskers *(barbels)* that flash and glow in assorted colors.

Prize for the most bizarre could go to female anglerfish. They have fishing poles that grow on top of their heads. Lighted lures resembling worms dangle from the ends of these poles. The anglers swish their bait back and forth. Should a lure be bitten off, a new one grows in its place.

Squid are among the most colorful of ocean dwellers. Not only are they able to change their skin colors, but they are true bluebloods. Their blood is blue. Deep sea squid light up like Broadway billboards. A one-yarder may have more than two-hundred colored lights on its body.

You will probably see some of these spineless beings in the Deep. Because they usually move through the water backward, many have bright silver taillights. When frightened, certain squid send out blobs of luminous bacteria ink about the same size and shape as their bodies—fiery ghosts to frighten enemies. Tiny deep-sea shrimp also send out luminous bacteria when frightened. With these "fireworks," they confuse enemies and escape.

Mini-monsters

According to our standards of beauty, deep sea fishes are ugly, repulsive, and monstrous looking.

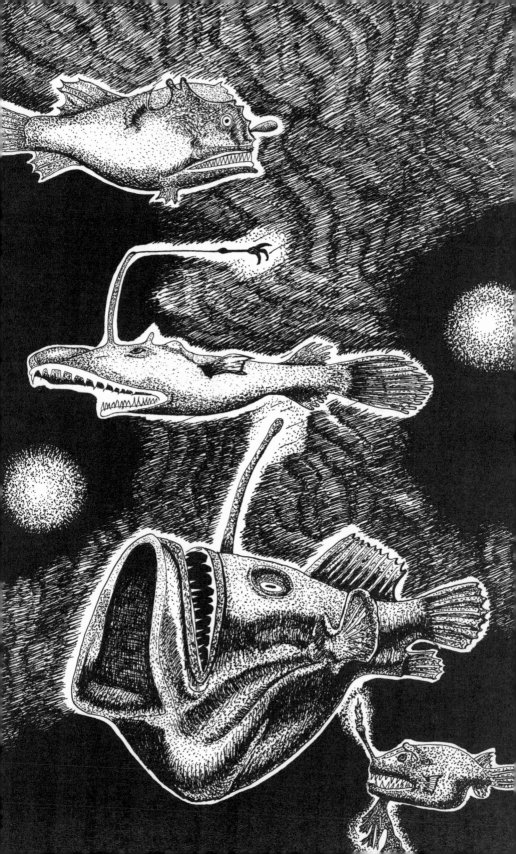

Most of them are not much bigger than your fist. They usually have needle-sharp teeth, huge jaws, and elastic stomachs. No bright orange hues or green stripes here! Most deep-sea fishes are dark brown or black.

It's common for fishes of the Deep to eat other fishes twice, even three times as big as themselves. Bellies stretch like rubber to make room for victims. The black swallower, for example, is only a few inches long. Yet it can swallow a fish measuring ten inches. In addition to black swallowers, deep sea perches, giant tails, and gulpers are belly stretchers.

The Bottom of the World

The *Dragon* descends almost five miles into the night world few humans have ever seen. It is a place of inky blackness and intense cold, with pressures measuring in tons per square inch. Like moths, many deep sea creatures will come right up to the Nose of the *Dragon* when the lights are on. Perhaps they welcome the brilliant entrance of the *Sea Dragon* into their world. Except for bacteria, which some experts classify as plants, no plant life exists in the sunless sea. It is remarkable that any form of life can survive in temperatures barely above the freezing point of seawater, and under pressures that can crush steel.

The enormous pressures don't squeeze the life out of animals of the Deep because their loose, flabby bodies have internal pressure that is the same as that of the surrounding water.

Any area deeper than twenty-thousand feet below the ocean's surface is called the Hadal Zone or Hellish Zone. A hot luncheon is served when you reach "Hades." The *Dragon* cruises close to the bottom while you dine, mouth agape. You carry your plate of food to the Nose, where you munch goodies while observing wonders.

As you cruise miles down, outside lights are beamed to the ocean floor. You see starfish running across the bottom on the tips of their arms. Hundreds of brittle stars, which are like starfish, link arms to form tangled carpets. Sea spiders scamper over the mud. Their bodies are so narrow that they carry most of their internal organs in bundles attached to their leg joints.

Claws and slurpers are put to use. Gil Haddock and Cliff Rockbottom are at the controls. They never know what they'll uncover. From the experience of other scientists they know that there are at least 310 species in the Hellish Zone. These include sponges, jellyfish, worms, snails, clams, sea urchins, and sea cucumbers, as well as assorted varieties of fishes.

Many creatures are blind. They have feelers enabling them to locate food and find their way about. Others, like hatchetfish, have enormous pop eyes keen enough to find the tiniest sparkle from a light-bearing prey.

Are there sixty-foot sea serpents slithering around

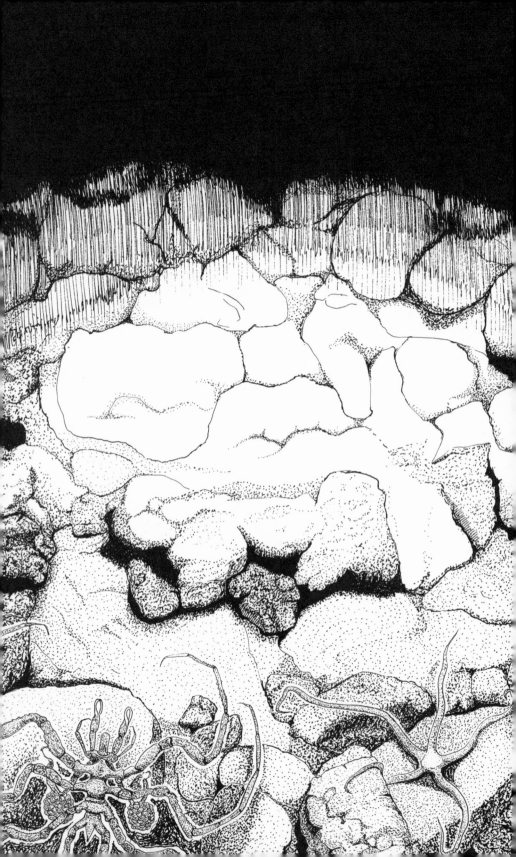

in the dark? Scientists are skeptical. But until one-hundred years ago the experts were positive that no life could exist more than one-thousand feet below.

Anything is possible, especially in the Hellish Zone.

VI. Visiting Fish Doctors' Offices

Fish have their own medical centers in tropical waters all over the world. These centers are known as *cleaning stations*. You visit one that is located just outside the Puerto Rican Trench. It's eighty feet down, under a cliff.

Sea Dragon passengers watch the busy swim-in office of the cleaner wrasse fish. At least twenty-six species of fish make cleaning their profession. Like them, wrasses advertise their business. They swing and sway in front of their offices. They sometimes make their skins turn gaudy colors as a come-on to attract customers. Business is usually brisk. Customers come from all directions and line up to receive treatment.

Groupers, huge sea bass, ocean sunfish, and even sharks queue with surgeonfish, parrot fish, and other mariners. Although many of these fish are natural enemies, they don't fight or bite here. The waiting room is quiet and orderly. Patients needing service behave patiently.

Cleaner wrasses give thorough examinations, going inside gills, and feeling safe even when they enter the mouths of giant fish to treat sore throats and clean teeth. They nibble away at dead flesh, parasites, and bacteria infecting wounds. Their pointed snouts act like tweezers as they get at imbedded foreign bodies.

Patients usually use body language to tell fish physicians about their problem spots. They may stand head-up-tail-down, lie on their sides, or turn upside down. Several change color to help the "doctors" see parasites and worn skin. Many open their mouths to indicate that their throats are sore or that their teeth need cleaning. Sharks and other big fish often have a team of "doctors" working over them all at the same time.

Scientist Conrad Limbaugh once counted three-hundred fish handled by a single cleaning establishment during a six-hour workday. To test their importance, Limbaugh removed cleaner fish from a Caribbean cleaning station. Within a few days customers stopped coming. Within two weeks fish in the area were in poor shape. Some had serious infections, open sores, and ragged fins. Others were sickly.

Phony imposter fish can ruin a cleaning station business. Wrasse blennies, resembling wrasses in color, shape, and size, sometimes swim with wrasses that clean other fish. Instead of helping patients, blennies bite off pieces of skin from trusting customers.

VII. Aboard a Y-Not

You have to board a mini-sub to tour shallow waters that the mighty *Sea Dragon* dares not tread.

Your *Dragon* surfaces near the Virgin Islands. A huge sub tender with a dozen Y-Not cub subs in its hold has been sent out to meet you. You watch cranes lift the Y-Nots into the water. Their walls and floors are glass.

These small subs are owned by the Y-Not Oil and Gas Company, which is also part owner of the *Sea Dragon*. The Y-Nots have been sent from the Texas Gulf Coast, where they are used to taxi technicians to and from oil rigs and pipelines.

The B-Cause Environmental Association opposes Y-Not. B-Causers protest their disturbing the living quarters of sea creatures, and they fear that oceans will become polluted by oil seeps and gas leaks. B-Causers on board are pleased that the Y-Not mini-subs have been diverted from drilling duty and, instead, are being used on this trip to educate people about the wonders of wildlife in the sea.

Each cub sub has room for ten passengers. It can cruise underwater at five knots (a little over five miles per hour) and is capable of descending a hundred feet while maintaining "one atmosphere"—normal sea level atmospheric pressure. It is able to hover, make sharp turns, go forward, backward, right, and left.

Walk the plank that takes you from the topside of the *Dragon* to the hatch of a Y-Not. Go down the hatch and sit down in one of the contour plastic chairs. Sally Forth, Gil Haddock, Cliff Rockbottom, and several other crew members from the *Sea Dragon* pilot the Y-Nots. The mini-subs sink approximately thirty feet under, heading toward the Virgin Islands. Within fifteen minutes you reach ancient undersea sculptures.

A City of Sea Creatures

You tour a stone city whose inhabitants are more varied and eccentric than any encountered by Alice in Wonderland or Dorothy in Oz.

The city's architects are creatures about the size of a nailhead. They are coral polyps, mini-monsters with six or more arms growing around their mouths. These animals make stone houses for themselves by taking calcium from seawater and changing it to limestone. When they die, their houses remain as a memorial to their skills. Some of these memorials have lasted millions of years.

Coral polyps have transformed many parts of our planet. They have created islands, added millions

of acres of land to continents, and built reefs that stretch for hundreds of miles.

Polyps are aided in their work by certain plants and shelled creatures who add cement and limestone to the structures.

The coral city you visit is still under construction. Millions of busy polyps living in the warm, sunlit waters of the Bahamas are building homes over those of other polyps who founded their city thousands of years ago.

Stone sculpture surrounds you. There are towers, globes, and irregular shapes of all kinds. Many look like trees with huge trunks and zigzagging branches. Others resemble stag horns, shrubs, giant fans, and mammoth tables. Coral formations can be every color of the rainbow. In this area they are chiefly yellow tipped with white. Tiny holes you see in the stones are polyp homes. Bright patches of red, green, orange, and blue decorating the stones are sponges and plants.

There's a greater variety of wildlife in a coral reef than in any other part of the ocean. How fascinating it is to observe the habits and lifestyles of animals from a glass Y-Not!

Many of the fish are quick-change artists. Goatfish, for example, can change their spots into stripes, and turn from white to rosy red.

Fishes seem to have reasons for their instant costume changes. In many cases, they color-coordinate with their surroundings. They may turn brown, for example, when resting on a brown sponge. In other instances they seem to whiten with fear, break out in yellow spots when ready to fight, and put on gaudy hues when eager to mate.

Gil Haddock, our ship's ichthyologist, likes to tell the story of a grouper that was blue with brown stripes. It swam into a coral cave and emerged a few minutes later dressed in yellow with black dots! Only an expert like Gil could know it was the same fish.

Groupers surely have odd ways. They also undergo sex changes. Like many other kinds of fish and shelled animals, they start life as females, and after a few years turn into males. Thus, they can be mothers and fathers in one lifetime.

You are bound to see three-foot groupers patrolling the coral city. They are among its largest inhabitants. Sharks, swordfish, barracudas, and other giants you may see near coral are unwelcome visitors who live elsewhere. These outsiders barge in for fast, self-service seafood dinners.

Fights are frequent among coral dwellers. Surgeonfish earned their name because they have knifesharp scales on the sides of their tails that snap open and shut like switchblades. Even those darling angelfish you see darting about have sharp spines that they use to slash and stab enemies. Triggerfish have unusual defenses. They raise a spine and lock it into place in order to wedge themselves in a coral cranny. Then they can't be pulled out by an enemy.

You'll see a variety of puffers. When attacked, they blow themselves up by gulping water. In that way they become hard to swallow.

Porcupine fish are puffers whose spiked skins have been turned into lanterns that decorate many sea-food restaurants. They cannot escape a tragic fate. They may be poisonous to eat, but they end up in restaurants anyway.

Poisoners are everywhere. The arms of coral polyps contain venom that they inject into microscopic animals that pass by. Then the victims are de-voured.

Sea anemones, like polyps, have arms with poison-ous stingers that grow around their mouths. Anemones are called flower-animals because they resemble beautiful chrysanthemums. Their ven-omous arms look like flower petals.

Box-shaped trunkfish are also resident poisoners. They can send venom that's strong enough to kill small-sized enemies. The trunkfish's cousins, cow-fish, are poison-boxes with "horns" above their eyes. Cowfish look simply bovine as they graze along coral shelves eating sea worms and shelled creatures.

Some grazers crunch chunks of coral to get at the plants and polyps inside. Parrot fish, easily seen because they come in assorted loud colors, have strong beaklike jaws. They bite into stone and grind it up with teeth growing inside their throats.

There's not an empty niche or cranny. Animals lurk in dark holes and beneath coral formations. Huge green moray eels hide in caves. Spiny black sea urchins cram together in crevices, sometimes in clusters of twenty. Snails, mussels, and tube worms dig their own tunnel retreats. Jawfish improve their homesites by blockading the entrances with bits of coral. They can even build fences, using their mouths to carry and stack bits of shells and stones.

There are some that seek protection inside the bodies of other animals. A pearlfish often lives in a sea cucumber's body cavity. You may see one of these eel-shaped fish peering out of the rear opening of a wart-covered blob that looks like a fat pickle. It's amazing that the cucumber is not bothered by its guest.

The mysteries of coral life are endless. Although the Y-Nots are small and cramped, you will want to taxi to and from the *Sea Dragon* often during the three days you spend at the Tongue of the Ocean.

Night Excursions

Take night trips. Only then do you see polyps and sea anemones bloom in all colors as they stretch out their poisonous arms. (During the day they lie hidden inside their stone rooms.) Sleeping fish hang motionless in water beds under coral ledges, their lidless eyes wide open. At night conches desert their hideouts to march about. Eyeless urchins move around waving and tapping their spikes in all directions. Light-shy lobsters, moray eels, and octopuses begin their night hunts, while butterfly fish, angelfish, and other swimmers active during the day hide and rest.

Certain parrot fish dress up before they bed down. Each one secretes a slimy nightshirt that covers it when it sleeps. Some experts believe that the slime repels enemies, and allows the parrot a safe night's rest. Parrot fish break out of their pajamas every morning, and take at least half an hour to manufacture new ones for themselves every time the sun goes down.

VIII. Floating Islands

After spending so much time inside, you are over-joyed when the hatch is opened and you can go outside to enjoy fresh air and the sight of the open sea. What a great feeling it is to walk the deck and gawk at the waves!

Your submarine surfaces in the middle of the Sargasso Sea. This is a sea within the ocean. It stretches more than halfway across the Atlantic and is about the size of the United States. The only land it touches is the island of Bermuda, which is inside its boundaries. It is bordered on every side by swirling currents of cold water. Yet the waters of the Sargasso Sea are warmer and saltier than the rest of the Atlantic.

Millions of tons of *sargassum* weeds float in the Sargasso. These are yellow and brown, with fringed leaves and pea-sized bladders. The weed was named by Columbus' sailors. The air-filled bladders that keep the weeds afloat reminded them of *salgazo*, the small grapes raised back home in Portugal.

Sailors used to believe that the weed masses stopped ships and ensnared them forever. They spoke of skeletons on rotting hulks that were entangled in masses of vegetation.

The Sargasso Sea was called "port of missing ships" and "graveyard of the Atlantic." Actually, the weeds aren't thick or heavy enough to trap a rowboat. Although they form clumps, the plants are widely scattered over fifty thousand square miles.

Using nets, the crew scoops up sargassum plants and dumps them into big tubs of seawater. The contents are then ladled into plastic bowls—one for each passenger. A different bowl is set before you each of the three nights you cruise the Sargasso.

Feast your eyes! You have before you wondrous globes containing bizarre miniature beings.

You feel like a fairy-tale sorcerer. Use chopsticks, or fingers, as magic wands. Grasp the weeds and shake them. Surprise! Each weed is a floating island inhabited by sea imps. They will fall into your bowl. These sea imps match the yellow-brown color of their vegetable-islands. Many are shaped to mimic parts of the plants, and can be found only in the Sargasso Sea. A magnifying glass has been set at each place. You need it to find the now-you-see-'em, now-you-don't animals.

The sargassum fish, world-famous master of disguise, looks like a ragged leaf. Its body is fringed from head to tail, and its blotchy colors blend with the weed. Tiny white dots all over its skin, even in its eyes, match the dots tube worms make on sargassum leaves. Because of its huge mouth, the sargassum fish is also called "frogfish." Like deep sea gulpers, its stomach stretches like a balloon. It qualifies as a member of the anglerfish family, for it dangles a fleshy lure from the top of its head to attract prey, including other sargassum fish. What a cannibal! Chef Saki Yaki once found six sargassum fish inside the stomach of a "big brother."

A sargassum fish can be the size of your thumbnail or as big as your fist. Notice its front fins, which bend like fingers. The fins grab stems and leaves, enabling the fish to crawl and swing like a monkey from weed to weed.

A raft of creatures survive on rafts of sargassum. Sea horses hook their curling tails around weeds. (Although some Pacific Ocean sea horses measure two feet long, the usual sargassum sea horse is less than an inch when full grown.) Using your magnifying glass, you may discover that the fringe on a plant is a one-inch octopus, and that the "grapes" are dime-sized jellyfish, sea slugs, and crabs.

The only insects that live in mid-ocean inhabit the Sargasso Sea. Water striders run across weeds on their six hairy legs. They are even able to run on top of water. The insects feed on drifting bodies of dead creatures. They lay their eggs on sargassum weeds, or on bubble floats made by tiny blind snails that drift upside down and occasionally shoot off purple ink.

Cradle of the Deep

The Sargasso Sea is a nursery used by fishes that travel here to mate, then leave after they lay their eggs. Eggs floating in the waters could be those of marlin, swordfish, and sailfish. When hatched, these giants-to-be are less than an inch long. Imagine landing a sailfish no bigger than a bee; a marlin the size of a fly; a swordfish two-thirds of an inch long, one-third of which is head. These babies may someday roam distant waters of the world stabbing and devouring huge enemies.

Flying fish make nests of weeds and lay long strings of eggs that are light enough to keep the plants from sinking to the bottom of the Sargasso Sea, which is two to three miles down. Newborn flying fish are yellow-brown to match sargassum. When old enough to leave their nursery they turn silver-blue. Then they're off, able to wing it out of water. You will undoubtedly see them zipping through the air several feet above the waves. When mature, many of them glide above water for as long as thirty seconds, at thirty miles per hour.

Newly hatched barnacles swim about in Sargasso waters. When they grow up, they glue themselves to the bottoms of ships and the bodies of fish. The youngsters swim about. As adults they are permanent hitchhikers that cannot move around. Mature barnacles have been described as oddballs that stand on their heads and kick food into their mouths with their legs.

Here is the most amazing true fish story: Both

American and European freshwater eels travel thousands of miles in order to lay their eggs in the Sargasso Sea. After the female lays from five to twenty million transparent eggs, and the male fertilizes them, both parents die. (You will probably not see the eggs, but you will see any number of young eels floating past. They look like strips of glass.) The tiny eels, called elvers, find their way back to European and American streams, brooks, ponds, and puddles. American eels take a year to reach their continental homes. European eels must undertake a journey that lasts two and a half to three years. They are pushed along during their twenty-five hundred-mile trip by the fast-moving currents of the Gulf Stream. (The Gulf Stream, the most famous of all currents, moves northward along the American coast, swirls through the North Atlantic, then heads toward Europe.)

The most fantastic part of the story is that although Old and New World kinds hatch in the same area, young elvers sort themselves out. One group takes the European trip. The other heads for America. Never has a single American eel been found in Europe, and no European eel has ever been seen in an American pond or stream.

IX. Enjoying the Mountains

It's time to push off and sink down for further adventure. The *Sea Dragon* dives beneath sargassum plants and heads east toward the islands of the Azores. Sonar screens show hills and then a huge flat plain that is three miles down.

After two days' travel, you're in the mountains. The submarine zigzags back and forth between jagged peaks called *seamounts* and strange-looking flat-topped mountains called *guyots*. Guyots are probably drowned islands.

Outside lights are beamed at the mountainsides. At times it's almost impossible to see anything, for there appears to be a snowstorm outside. The "snow" is made up of skeletons, waste, and decaying matter that rain down from the upper regions of the ocean. In addition, dust in the air, sands carried by the winds, even specks of meteors from Outer Space drift down to rest on the ocean floor.

Parts of the Atlantic have carpets of "snow" that are two miles thick on the bottom. (People used to believe that many objects floated forever without falling to the floor of the sea. Everything eventually falls to the bottom, even beer bottles, tin cans, and other human-made trash.)

Cameras shoot large, clear panoramas. As soon as the photographs are taken they appear on several screens.

You are in the middle of the Mid-Atlantic Ridge, a mountain range that occupies about one-third of the Atlantic Ocean. It runs twelve-thousand miles, the length of the ocean from Iceland to Antarctica, and it's from three to twelve-hundred-miles wide. The Mid-Atlantic Ridge is part of a forty-thousand-mile mountain range that snakes through every ocean on earth. The entire range, called the Mid-Oceanic Ridge, winds itself around our planet like seams on a baseball. Greater than the Rockies, the Andes, and the Himalayas, it covers as much of the earth as all continents combined.

Most of the mountain peaks are a mile or more below the waves. Some of these are over twenty-thousand feet high. A few poke out above the water as islands. Iceland and the Azores are examples. Islands like these and mountains under the sea were formed by volcanic eruptions in the sea. They were made of lava that flowed from an inner part of our planet.

New lands are still being created. After an undersea

eruption in 1963, the North Atlantic island of Surtsey appeared overnight.

New islands can also disappear with mysterious suddenness. When a new Icelandic island was discovered in 1783, the Danish government sent an expedition to hoist the Danish flag on the new territory. The island couldn't be found. Like many newly emerged lands made up of volcanic ash, it was washed away by ocean waves.

Spending time in the mountains is very exciting, especially for Cliff Rockbottom, who hopes to write an original paper about the deepest part of the ocean. Claws and slurpers are working night and day. Cliff and his volunteer aides know that ooze from the ocean bottom may be worth a hundred times its weight in gold though it looks like a mess of mud.

Chunks from hillsides and muck from valley floors will be locked inside one of Purser Moneysworth's vaults. They are treated with the importance given to moon rocks. With the aid of other scientists, Cliff will take months, perhaps years, to analyze the rocks and bottom ooze for their age, origin, and chemical content.

Into the Valley

You descend into the Rift Valley, a vast V-shaped canyon that splits the entire Mid-Atlantic Ridge in the middle. In some places the Rift is thirty miles wide.

The *Sea Dragon* cruises at a leisurely pace, eight-thousand feet down. It circles around Mt. Venus (a tribute to the Roman goddess of love) and then winds around nearby Mt. Pluto (named after the Roman god who ruled Hades).

In the Hadal Zone, near Venus and Pluto! The experience always inspires Mme. Mia Solo to sing songs about love and doom. Whenever she performs, Commander Down's eyes water. He is her most ardent admirer. (Rumor has it that he has arranged for Mme. Solo to be a guest on every voyage.)

The mechanical claws dig into Venus and Pluto. They slurp in samples of black, glassy rocks. These rocks were once lava that flowed out of the earth. Each passenger receives a chunk of the black glass as a souvenir.

As the *Sea Dragon* snakes around the foothills of the mountain range, Rockbottom is aided by trailing TV cameras as he examines the sea floor for hot flowing lava. Tiny eruptions and sea quakes constantly take place in the Rift Valley. So far, Rockbottom has never seen a large open volcanic crater. But he has found twelve-foot-high "haystacks" of lava that resemble termite hills. Lava oozes from the top like toothpaste squeezed from a tube.

Scientists claim that lava not only makes mountains and islands, but also it creates additional sea floor that pushes America away from

Europe and Africa at the rate of an inch or two each year. According to the most recent theories, the ocean floor moves outward from the Rift Valley, pushing continents apart and making the Atlantic Ocean wider.

The Y-Not Company and other mining enterprises have a special interest in the bottom of the sea. They expect to mine coal in the waters of New England, gold off the shores of Alaska, and diamonds from the seabeds of Southwest Africa.

Millions of square miles of the ocean floor are paved with small rocks that look like burned baked potatoes. Industry considers these rocks, called *nodules*, to be the true treasures of the deep. They contain valuable metals, such as manganese, nickel, copper, cobalt, and iron. Mining companies are spending millions to design equipment that can collect nodules by the ton. You will see nodules in the Rift Valley. Just as scientists don't know how these were formed, engineers still don't know how to collect them in huge quantities.

The Rift Valley is also supposed to be a likely place to find hot springs, called *thermal vents*. Hot water vents have been found in the valleys of the Pacific and in the Red Sea. Scientists believe that they also exist in the Atlantic Rift. We'll see when we get there.

Sonar equipment and cameras towed beneath the submarine help examine the seabed. Thermometers sense temperature changes. Boiling water can spurt out of vents at temperatures as high as 660 degrees Fahrenheit, hot enough to melt lead.

Cliff and his helpers work feverishly. Mud around hot springs is rich in copper, zinc, iron, gold, and silver. Gil Haddock watches with bated breath, because several scientists have seen new species of animals living near hot vents.

These newly discovered creatures use sulfur as a source of energy. Green plants, which depend upon sunlight, are part of the food chain for other living things. But in the world of deep sea vents, strange animals thrive upon bacteria, which are nourished by chemicals that gush out of the earth.

Perhaps you'll find clams the size of dinner plates, animals that look like spaghetti, and six-foot worms with red-plumed heads. These animals were discovered near spurting hot springs at the bottom of the Pacific Ocean. There's a good chance that they cluster around hot vents here, also.

Should the *Sea Dragon* fail to find itself in hot water, be consoled by the fact that you will once again see strange animals that fascinated you in the Puerto Rican Trench; the gulpers, the lighted fantastic ones, and the stomach stretchers. Skates, sea cucumbers, starfish, and brittle stars decorate the floor. Should one of the ship's claws disturb a brittle star, the creature will break apart. All its arms will fall off—no tragedy in this magical sea. Within a short time the brittle star grows new ones and is once again a five-armed wonder.

X. Sea Chase

Up from the valley, over the mountains, and onward across the sea, Spywater sets the course. Prepare to sink, surface, circle, head north, veer south, and backtrack in your pursuit of wildlife. Eventually you head east to Portugal.

You will, undoubtedly, come across Portuguese man-of-wars. Fleets of one hundred are often found bobbing up and down on the waves. They look like football-shaped blue balloons. Although harmless looking, they are killers.

When Spywater steers the *Sea Dragon* into their midst go down to the Nose. You will see the monsters' many trailing threadlike tentacles. Some of these can be seventy feet long, about equal in length to a string that dangles to the ground from the top of a seven-story building. Each of the man-of-wars' threads contains poison similar to that of cobra snakes. Fish who blunder into the poison threads become paralyzed. Their lifeless bodies are then lifted by these threads and placed in the "balloon's" gaping mouth.

Despite the deadly venom, shepherd fish choose to live within the curtains of man-of-war tentacles. Either the shepherds are immune to the poison, or they are able to avoid touching the killer threads. At any rate, these three-inch fish manage to survive. Their touch-and-die curtain protects them from outside enemies, and they enjoy leftover crumbs of food dropped from the mouth of their Portuguese sea lord.

Small-fry amberjacks also use man-of-war tentacles as a protective cage. When full grown they slip away—very carefully—from a touchy situation.

Spywater can't predict what he'll find. The sea is still a deep, dark mystery, hard to fathom. Most of the ocean has never been explored. And little is known about travel habits of most of its inhabitants.

Looking for Migrators
Many kinds of fish, shrimp, squid, and jellyfish travel not only to and fro, but up and down. When night falls they come up from lower depths to feed on plants and animals on the surface. They sink again as the sun rises. The creatures use the black of night as a cover against enemies. They don't make nightly up-down migrations when the moon is bright. However, once on top, they are often attracted to artificial lights.

Spywater shines beams in all directions. Flying fish often leap on deck, and squid embrace the *Sea Dragon*'s cameras. Don't be surprised if you see deep sea creatures like hatchetfish, lantern fish, and bristlemouths (whose mouths bristle with teeth). They spend their evenings up top, after journeying from their homes six-hundred to twelve-hundred feet down.

It is amazing that these up-down migrators are able to withstand the change in pressure and temperature. They can leave waters fifty times heavier and thirty degrees colder and adjust to lighter, warmer waters above every night.

Land Ho!

The last day of the trip! The last lap of waters! As you approach Europe, Spywater sinks the *Sea Dragon* a mile down so that you can watch the ship climb up and over the *continental slope* to the continental shelf. The shelf, which borders the coast, gradually goes uphill to shore from a depth of about five-hundred feet.

Most of the world's sea creatures make their homes along the continental shelves. That's because shallower water has more *plankton*, tiny plants and animals that drift near the surface. Plankton make up the ocean's meadows. Herds of fish browse in these meadows, followed by fleets of fishermen, who hope for good net results.

Clusters of eggs and the young from all kinds of sea creatures are found in plankton. Immature fish, sea urchins, snails, starfish, crabs, lobsters, and barnacles are among those of the younger set.

Many plankton dwellers migrate up and down. They live down under during the day, and come up for the nightlife.

Chef Saki Yaki attaches fine strainers to kitchen slurpers in order to collect plankton. He concocts a wonderful soup that is tasty because of the vegetables, eggs, fish, and shelled creatures in the plankton. Plankton soup is the specialty for the Captain's Dinner, a traditional celebration enjoyed the night before you dock.

Passengers disembark at Lisbon. After remaining in port for two days, the *Sea Dragon* embarks on a return trip over the ocean and under the waves. Back to Norfolk, idling over the Mid-Atlantic Ridge, the Sargasso Sea, and exploring the Caribbean. The same areas, but the landscape and wildlife are always different. In the ocean world it's impossible to examine the same water twice.

BIBLIOGRAPHY

Baker, Robin. *The Mystery of Migration*. New York: Viking Press, 1981.

Blond, Georges. *The Great Migrations*. New York: Macmillan, 1956.

Brindze, Ruth. *The Sea*. New York: Harcourt Brace Jovanovich, 1971.

Clarke, Arthur C. *The Challenge of the Sea*. New York: Holt, Rinehart & Winston, 1960.

Coombs, Charles. *Deep Sea World*. New York: William Morrow, 1966.

Cromie, William. *Exploring the Secrets of the Sea*. Englewood Cliffs, New Jersey: Prentice-Hall, 1962.

Editors of Encyclopaedia Brittanica. *The Ocean: Mankind's Last Frontier*. New York: Bantam/Brittanica Books, 1978.

Engel, Leonard, and editors of Time-Life. *The Sea*. New York: Time-Life, 1969.

Gaskell, T.F. *The Gulf Stream*. New York: Doubleday, 1973.

Grzimek, Bernard. *Animal Encyclopedia*. New York: Van Nostrand Reinhold, 1972.

Hardy, Sir Alister. *The Ocean Sea*. Boston: Houghton Mifflin, 1956.

Idyll, C.P. *Abyss: The Deep Sea and the Creatures That Live in It*. New York: Crowell, 1976.

McNally, Robert. *So Remorseless a Havoc*. Boston: Little, Brown, 1981.

Miller, Robert. *The Sea*. New York: Random House, 1966.

The Ocean Realm. Washington, D.C.: National Geographic Society, 1978.

Ommaney, F.D., and editors of Time-Life. *The Fishes*. New York: Time-Life, 1969.

Ricard, Matthieu. *The Mystery of Animal Migration*. New York: Hill & Wang, 1968.

Shepard, Frances P. *The Earth Beneath the Sea*. Baltimore: Johns Hopkins, 1967.

Silverberg, Robert. *The World of Coral*. New York: Duell, Sloan & Pearce, 1965.

———. *The World of the Ocean Depths*. New York: Meredith, 1968.

Soule, Gardner. *The Greatest Depths*. Philadelphia: Macrae Smith, 1970.

———. *The Wide Ocean*. Chicago: Rand McNally, 1970.

———. *Undersea Frontiers*. Chicago: Rand McNally, 1968.

Street, Philip. *Animal Migration and Navigation*. New York: Scribner's, 1976.

Teal, John, and Mildred Teal. *The Sargasso Sea*. Boston: Little, Brown, 1975.

INDEX